Handwritten note at top of page:

Gross Vehicle Weight 817 kg
Curb Weight 665kg - 265 kg of Batter...
Weight of car without batteries 400 kg.

Welcome!

Welcome! 1
An Introduction to the Electric Car 3
G-Wiz! The little electric car that started a revolution 7
Living with a G-Wiz 9
Will a G-Wiz work for me? 15
Me and my G-Wiz 23
Purchasing and Running Costs 27
Electric cars and the environment 29
Real world economy figures 43
A final word 47

This is a book about the most popular electric vehicle available in the UK today – the fun, quirky and unmistakeable G-Wiz.

Love it or loath it, the G-Wiz is a revolutionary vehicle. It runs on electricity and not fossil fuels. It emits no pollution and travels in silence. In short, it's a car that rewrites the rulebook.

This book is for anyone who is interested in owning a G-Wiz. If you want to know what these cars are like to own, use and live with on a day to day basis, this book will put you in the picture.

I've been involved in the electric vehicle industry since 2003 and I've been using a G-Wiz on a daily basis as my own personal transport since early 2006.

As well as owning and using a G-Wiz for almost five years, I run the REVA and G-Wiz Electric Car Club, the largest electric car club in Europe with over 1,000 members.

Over the past five years, I have conducted numerous tests with the G-Wiz. These range from formal tests on the environmental benefits, to the simple daily test of seeing how well it works as a family car with my two young children.

I've also carried out an extensive survey, covering both existing electric vehicle owners and the wider car-owning community, to find out the true story about electric car ownership.

This guide is based on my other electric car book, *The Electric Car Guide.* I have simplified and updated it specifically for people considering the purchase of a G-Wiz.

By the time you have finished reading this book you'll understand what it is like to own, use and live with one of these entertaining cars on a daily basis. You'll understand the benefits and the drawbacks and you'll know whether or not a G-Wiz is a suitable vehicle for you.

Greenstream Publishing
12 Poplar Grove, Ryton on Dunsmore, Warwickshire, CV8 3QE
United Kingdom

www.greenstreampublishing.com

Published by Greenstream Publishing
2010
Copyright © Michael Boxwell 2010
ISBN 978-1-907670-05-3
Michael Boxwell asserts the moral right to be identified as the author of this work.

An Introduction to the Electric Car

It seems every week electric cars are hitting the headlines. From car makers showing their latest concepts, governments announcing new incentive schemes and announcements of new technology breakthroughs, electric cars appear to be everywhere. Yet when was the last time you saw an electric car on the road?

The chances are you may never have seen one at all. Unless you live in the centre of London, Paris or Bangalore, the three 'electric car capitals' of the world, electric cars in actual daily use are few and far between.

Electric cars are nothing new. The first electric cars were built almost 175 years ago. At the beginning of the 20th Century, electric cars were the best selling cars around. Owners liked them because they were smooth, quiet and easy to drive. They only fell out of favour because internal combustion engine cars became cheaper to buy and fuel became more readily available.

It took a tiny Indian car manufacturer to bring electric cars back into the public eye. The REVA Electric Car Company launched their first vehicle, simply called REVA, in 2001.

After initial success in India, a European version made its debut in the United Kingdom in 2003. Branded G-Wiz for the UK market, it was initially offered in Leeds followed by London.

The success of the G-Wiz has played an important part in the reintroduction of electric cars by the major car manufacturers. Public acceptance of the tiny car has shown them that there is a strong market for electric cars. Policymakers have further encouraged the uptake of electric cars with the introduction of electric car charging points and electric-car friendly policies.

As a consequence, almost every car maker now has active electric car development projects, with the majority of these cars scheduled to appear between now and 2015.

Yet the charm and the abilities of the G-Wiz are undiminished. Today it is sold throughout the world with dealers in 22 countries. It is the most successful road going electric car in the world.

Why buy an electric car?

Buying an electric car isn't just about the environment, or worries about oil reserves. As G-Wiz owners have found out for themselves, these cars are fun to drive. Sprightly around town performance, an incredible turning circle and ease of use give the car some significant benefits.

It is no surprise that so many people don't want to go back to driving a petrol or diesel powered car after experiencing an electric car for themselves.

What is an electric car?

An electric car is powered by an electric motor, typically powered using energy stored in a battery. It is charged by plugging it into a power socket.

Like most electric cars, the G-Wiz does not use a conventional gearbox as the motor has a more flexible power delivery than an internal combustion engine.

Electrical energy is stored in a bank of batteries built into the car. In the G-Wiz, these batteries are fitted under the floor, giving the car a low centre of gravity.

Charging the car from a 240v domestic socket takes around eight hours. A top up charge to 80% of capacity takes 2½ hours allowing the G-Wiz to be 'topped up' between journeys.

How the G-Wiz compares to a conventional car

While the G-Wiz is not for everyone or every application, it is an ideal vehicle for many people and offers significant benefits over other types of cars.

At the heart of the G-Wiz is the electric motor. In terms of construction and power delivery it is almost the complete opposite of an internal combustion engine:

	Electric Motor	Combustion Engine
Average Efficiency	80% plus	25-35%
Maximum Power	From Standstill	At high speed
Multi-ratio gearbox required	No	Yes
Number of moving parts	3	130+

Because of these different characteristics, driving a G-Wiz is different to driving any other car. Maximum power from a standing start means it pulls away smoothly. You don't have to rev the engine to pull away. There is no vibration from the motor and they are exceptionally quiet.

Because of the power delivery of the electric motor, low speed acceleration is instant whenever you need it. When you take your foot off the accelerator, the motor is used to slow the vehicle down and recharge the batteries at the same time. With early G-Wiz models, this took a little getting used to as the braking felt very different from other cars. Later models have an improved braking system. The result is much smoother, more progressive braking plus a smoother transition between accelerating and braking.

The G-Wiz is very easy to drive in heavy stop-start traffic. There is no gearbox or clutch to worry about and the car can crawl along at low speeds very efficiently with minimal effort on the part of the driver.

Several people who suffer from travel sickness when travelling in a conventional car tell me they experience no problems when riding in a G-Wiz.

The G-Wiz is not a fast car. It has been designed for city use and has performance to match. This is ideal in the ebb and flow of traffic in a busy city. Later versions of the car have improved performance making them also suitable for shorter trips driving on open roads.

The G-Wiz is in its element in built up areas such as cities, urban and sub-urban environments where speeds are limited and there is a lot of stop-start driving. The size and incredible turning circle also help in built up areas. You'll nearly always be able to find somewhere to park in a G-Wiz!

Range

Range is always an emotive subject when discussing electric cars. Wherever you go around the world, it is the number one concern that non-electric car owners have about owning an electric car, with the perception that they could never adapt to a car with a range of 'only' 40 miles.

The reality is very different from the perception. Many G-Wiz owners will actually describe the freedom they feel that every time they go out to their cars in the morning. They know they have enough 'fuel' to go wherever they want to without the hassle and cost of visiting the service station.

One electric car owner explained it to me like this: "It takes me nine seconds to charge up my electric car. That's the time it takes to plug the car in when I get home. The next time I need to use the car, it is charged up and ready to go."

There are four different models in the G-Wiz line-up, and the official range varies from 40 miles (65km) on the early cars to 70 miles (112km) on the latest G-Wiz L-ion, fitted with lithium batteries (these different models are described in more detail in the next chapter).

Realistically, range will vary depending on driving style and conditions. The official range is based on careful driving in reasonable conditions. Drive flat out wherever you go and the range will drop significantly. Drive in extreme cold conditions in the middle of winter and likewise, the range will be lower.

According to the UK Department for Transport, the average car journey is 6½ miles (10½km) with 93% of all car journeys being less than 25 miles (40km). All of these average journeys are comfortably within the range of the G-Wiz.

It is interesting to compare the concerns that non-electric car owners have about range with the perceptions that existing electric car owners have about range:

- Non-electric car owners perceive that range is going to be a constant issue. They believe that they will be restricted because they cannot simply visit a filling station to refuel their cars.
- Electric car owners like the fact that every time they go out to use their car it is fully charged up and ready to go. They have enough fuel to go wherever they need to and they'll never have to visit a filling station ever again.

Of course, range is one of the big differences about owning an electric car and I cover this subject in much greater detail in a later chapter.

Chapter Summary

- Electric car ownership is not just about the environment or worries about oil reserves.
- They can be fun to drive, with ultra-smooth power delivery and are easy to use.
- Charging time from a 240v domestic power socket takes around eight hours.
- You can recharge to 80% of the battery capacity in around 2½ hours.
- Thanks to the flexibility of the electric motor, the G-Wiz does not need a gearbox.
- Power delivery is smooth, making for a more relaxed driving environment.
- Range is always seen as a big issue. The reality is different from the perception. Electric car owners appreciate the benefits of recharging a car from home rather than refuelling.

G-Wiz! The little electric car that started a revolution

In early 2004, a tiny new car appeared on the streets of London, confusing and surprising everyone who saw it. It was more compact than a Smart. It had cheeky, quirky styling that demanded attention. It could turn around in the width of a narrow street, squeezed into the tiniest parking spaces and travelled in silence.

The car attracted attention wherever it went. What was it? Who made it? Why is it so quiet? Was it electric? Suddenly, the media were falling over each other to report on a new type of vehicle: the G-Wiz.

In a remarkably short period of time, the little G-Wiz went from being a complete unknown to a media sensation. TV celebrities, Hollywood stars, politicians and captains of industry were queuing up to buy them; fashion designers created special editions. Suddenly, the G-Wiz became cool.

Hundreds of people flocked to buy G-Wiz's. Some were attracted by the cheap running costs, free parking and exemption from London's Congestion Charge tax. Others were attracted by its environmental credentials. Lots were bought by owners of large four-wheel-drive cars, wanting a smaller vehicle to use in the city.

Local government responded by installing charging points and offering free parking in many parts of the city. Enthusiastic owners went on to form the G-Wiz Owners' Club – now one of the largest electric car owners clubs in the world, and the EV Network – the world's first National Charging Network for electric cars.

Other car makers watched in astonishment: why were people buying a small electric car over their own far superior models? Why was there a five month waiting list for the G-Wiz? It was the first indication that the general public was eager to buy electric cars.

Honda, Renault and Peugeot showed electric concept cars of similar dimensions to the G-Wiz. Norbert Reithofer, CEO at BMW started talking about BMW building a G-Wiz competitor, explaining that this will be an important future market for BMW.

Few motoring journalists understand the G-Wiz, nor understand why it has been successful. Few cars have divided opinion or created as much debate as the G-Wiz. Few cars are likely ever to do so again.

Today, over seven years after London saw its first G-Wiz, the car remains popular. Visit a car park in the City of London or in Westminster and you'll see more G-Wiz's than any other single model of car. Many Londoners view the car with pride – it's 'their' car, created for their city.

A recent Frost and Sullivan survey into electric vehicles noted that when people find out about electric cars, their interest in buying one doubles. I carried out a survey with city car users as research for this book. In cities where the G-Wiz is common, people are twice as likely to consider buying an electric car in the future.

Each day, the G-Wiz demonstrates to millions of Londoners that electric cars are practical. It's the first time any electric car has done so in 100 years. In doing so it has become a milestone in the uptake of electric cars and a blueprint for the future.

It is a legacy that will live on for a very long time to come.

Living with a G-Wiz

The G-Wiz is a compact, two door city car with space for two adults and two small children. It is designed for use in built up areas for short journeys.

It has a high seating position, giving occupants excellent all-round vision and making it easy to get into and out of the front seats. Inside, the switchgear is robust and ergonomically placed and it is easy to get a comfortable driving position.

Different G-Wiz models

Since its launch, there have been four different models of G-Wiz. Each one has been a significant step forward over previous models, although the basic shape of the car remains the same.

The first car, now called the G-Wiz dc-drive, had a top speed of 40mph (65km/h) and a range of up to 40 miles (65km). This was sold in the UK between autumn 2003 and autumn 2006 and remains the best selling model in the range.

The second generation, called the G-Wiz ac-drive, had an improved top speed of 48mph (77km/h), better acceleration, improved handling and braking and a range of up to 48 miles. The car sold between autumn 2006 and the end of 2007.

At the end of 2007, the third generation G-Wiz was launched: the G-Wiz i. A redesigned interior increased space and comfort for front seat passengers. Top speed and acceleration had improved again with a top speed of 51mph (82km/h), handling and braking was improved thanks to input from Lotus and the car had improved crash protection with a new safety cell.

In 2009, the G-Wiz i was joined by a longer range model, the G-Wiz L-ion. With lighter, more powerful batteries, the G-Wiz L-ion had a longer range of up to 70 miles (112km) and could be recharged from flat to full in less than six hours. The G-Wiz L-ion could be fast charged from a specific charging unit in 90 minutes.

First impressions

There is no doubt that the G-Wiz provokes a reaction when you first see the car. Some people love it, some people hate it, but everybody has an opinion. Quirky, charming and cute? Or ugly, basic and crude? It depends on your point of view, but there is one thing that everyone can agree on: the G-Wiz makes an immediate impact.

Get in the car, drive off and the car makes even more of an impression. The silence, the smoothness and speed off the line take bystanders by surprise. Everyone's first reaction is always 'Wow!' and the car is often dubbed 'cool' by teenagers.

Once you switch on the ignition, there is no noise or vibration to let you know the engine is running. Most people are unsure whether the car is actually ready to pull away or not. As you tentatively put the car into Drive and touch the accelerator, there is a mild sense of shock that the car pulls away virtually silently.

It takes a few minutes to get used to the strange sensation of travelling without an engine noise. Some models of the G-Wiz have a motor fan that runs when the car is moving, but at speeds below 10mph, the car makes very little noise and as a driver you are aware of that.

As the speeds increase, road noise and wind noise now mean that your electric car is making virtually the same amount of noise to the outside world as any other car.

Performance

Once you are used to the silence, you start noticing other things. The power delivery is very smooth and there is no vibration from the engine.

The G-Wiz cannot be described as a fast vehicle, but performance is fine around town where acceleration is reasonable. The early models of G-Wiz (the dc-drive version sold before 2007) struggle on hill-climbing. Later versions are significantly better, also providing significantly better acceleration and higher top speed.

Likewise, earlier models of G-Wiz do feel under-powered on open roads. The later vehicles are significantly better for driving outside of towns.

The Fun factor

Most people who have never driven an electric car are quite surprised by how much fun a G-Wiz can be.

It won't win a race away from the traffic lights against a Porsche. Yet in their natural environment of a town or a city, G-Wiz's are surprisingly nippy. The spritely performance and responsive steering make for an entertaining and enjoyable driving experience.

Braking

Braking for the first time can be strange experience. The G-Wiz has two braking systems, combining 'regenerative braking' with a standard braking system. Regenerative braking uses the momentum of the G-Wiz to generate power which is put back into the batteries. This can improve the range of the car by as much as 30%.

To slow down, all you do is put your foot on the brake pedal and all the power is then fed back into the batteries to extend your range. Push on the brake pedal harder and the mechanical brakes kick in to help slow the car quicker.

In reality, unless you have to stop quickly or if you brake hard all the time, you are unlikely to be using the mechanical brakes that often: the regenerative braking is sufficiently powerful to stop the car in normal day to day driving.

Range Fixation

It isn't long on your first drive before you become aware of the fuel gauge. In fact, if you've never driven an electric car before, the fuel gauge almost always becomes a fixation for the first few weeks.

Like every electric car, the G-Wiz has a shorter range than a petrol or diesel powered car, so the fuel gauge moves quicker than you will first expect.

In most cars the fuel gauge seems to stay close to full for the first half tank of fuel and then moves down quite rapidly afterwards. In any car, once the fuel gauge drops below a quarter, many people start getting 'range fixation'. They're on the lookout for a fuel station and starting to worry if they can't find one on their route.

In a G-Wiz, the fuel gauge will start to move after only a mile or so of driving. It is a bit disconcerting at first because everyone is so used to the way fuel gauges work in conventional cars. In effect, you are getting the same range fixation as you do when running low on fuel in any other car.

Even when you're driving a short distance and you absolutely *know* there is enough charge to get to your destination, it is very easy to get fixated on the fuel gauge in the early days.

It's a psychological difference. In reality, you are no more likely to run out of range in an electric car than you are to run out of fuel in a normal car. The best analogy to use is that of a mobile phone. If you plug your phone in overnight to charge it up, it doesn't let you down. The same is true with an electric car.

Once you've got more confidence in your G-Wiz and you have used it for a while, your range fixation disappears.

In fact, you get to the point where you ignore the fuel gauge completely. After all, if you plug your car in every night and you know that you've got enough range to do your daily driving, why bother checking the fuel gauge?

Range Fixation with a brand new G-Wiz

The issue of range fixation is worse on a brand new electric car than with a used car.

If you have a brand new G-Wiz, the batteries will need to be used a few times to build up the range. You won't get your maximum range until the batteries have been run in by going through a few charge-discharge cycles. When the car leaves the showroom, it may only do around two-thirds of its advertised range. If you are not aware of this problem, this compounds the problem of range fixation. It certainly doesn't do much for confidence in your new purchase!

The range will improve within a few days and continue to improve over the following weeks. Within a few weeks, your batteries will be run in and your range will be significantly better.

Plugging It In

It feels very strange plugging a G-Wiz into a power socket for the first time. It is a novelty that takes some time to wear off. Yet, very quickly it becomes a convenience: no more do you need to travel to a service station to refuel your car. You simply do it when you get home.

The G-Wiz has a power socket hidden behind a fuel filler flap on the rear wing. A power cable connects the car to a standard household power socket. An external weather proof power socket can be fitted to the outside wall of your home for convenience.

The G-Wiz takes eight hours to complete a full charge from flat to full from a domestic 220v power socket. A quick 'top up' charge can take the car to 80% of capacity in around 2½ hours – ideal for topping up the charge between longer journeys.

The first few weeks

The first few weeks with a G-Wiz are fun. The novelty factor of a car that runs on electricity lasts a while and the G-Wiz is a lot of fun to drive around town.

New owners go through a period where they are thinking about every journey they go on. They constantly check they have enough range and often make arrangements to plug their car into a power socket at their destination, even if it is well within the range of the car.

It is all part of the 'range fixation' that new electric car owners have. It soon wears off.

One thing that happens to most people at some point in the first few weeks is forgetting to plug the car in. Most people do it once – and usually at the most inconvenient time! Thankfully, if your journey is relatively short, you can often plug your car in and get enough range after 20-30 minutes of charging to get you to your destination. As a consequence, the results are rarely catastrophic but most electric car owners rarely make the same mistake twice!

Speed and Range

As with any other car, economy figures depend on how you drive the car. If you drive everywhere as fast as possible, you will not get the same range as you will if you drive economically.

Performance and range depends on which G-Wiz model you have:

	G-Wiz dc-drive	G-Wiz ac-drive	G-Wiz i	G-Wiz L-Ion
Top Speed	40mph 65km/h	48mph 77km/h	50mph 80km/h	50mph 80km/h
Range	30-40 miles 48-65km	36-48 miles 57-77km	38-48 miles 60-77km	50-75 miles 80-120km

There are many factors that do make a difference to the range in a G-Wiz. Of course, running heating or air conditioning will reduce the range. Less obviously, ambient temperature can also make a difference. Batteries perform better in warm weather than they do in very cold conditions.

Using lights or radio in the car will make very little difference to range. These ancillaries use relatively insignificant amounts of electricity compared to the amount of energy used by the electric motor.

All car manufacturers are wary of claiming unfeasibly long ranges for their cars, as to do so would damage customer confidence in their products. Yet it is also true that to achieve the maximum range, some drivers have to adjust their driving styles and techniques in order to achieve them. These adjustments are not difficult and many people adopt them without even being aware that they have done so.

During the first few weeks, most drivers experiment with different driving styles to see how much they can improve the range of their cars. Even if they don't need the maximum range from the car, many drivers feel a sense of achievement by getting the absolute maximum range possible out of their G-Wiz.

As with any type of car, the biggest single difference you can make to improve your economy is adjust your speed. The faster you go, the greater the wind resistance and the shorter distance you'll be able to drive. Conversely, the slower you travel, the further you'll be able to go.

The calculation is not linear but as a general rule if you reduce your speed by 10% you'll be able to increase your range by around 15%.

Another good example of how a driver may adjust their driving techniques with a G-Wiz is with braking:

- In conventional cars, a huge amount of kinetic energy is lost when you apply the brake pedal. The energy is converted to heat through brake friction.
- In a G-Wiz, regenerative braking uses the speed of the vehicle to power the motor, which in turn generates electricity that charges the batteries, running the entire system in reverse to generate electricity.
- With a little practice, regenerative braking can handle most of the braking effort required in day-to-day driving.
- In a city environment, using regenerative braking effectively can increase the range the G-Wiz by up to 30%.

Naturally, it takes time for a new electric vehicle driver to get used to regenerative braking and to learn how to use it as effectively as possible.

Freedom from the service station

After driving a G-Wiz for a few weeks, you see that every time you get up in the morning you have a car with a 'tank full' of electricity and the freedom to drive wherever you want during that day.

Suddenly, the benefits of being able to recharge your car at home rather than having to drive to a service station to refuel become apparent. No longer are you worrying about range; rather you are seeing the benefits of always having a car with a tank full of electricity every morning and never having to pay for fuel at a service station ever again.

Borrowing a 'plug full' of electricity

At some point during your electric car ownership, you will ask a friend if you can charge up your car when you're visiting. Most people are more than happy to 'lend a plug'. When people know that you are driving an electric car, many people will offer without being asked. Friends are often quite surprised if you don't need the charge and turn them down!

If you are in the UK, the cost for the electricity your borrowing is likely to be in the region of 25p per hour during peak times until the car batteries are 80% full, at which point the charge rate drops and the cost drops to under 10p per hour.

If you use your electric car for business, you will often find that businesses are also more than happy to offer a charge up when you are visiting them. It is always best to phone up and ask first, to make sure that it is convenient.

If you are visiting a remote pub, restaurant or even an independent hotel in the evening, you will find that most of them are more than happy to offer a charge up in return for your custom. Always phone up and ask first though, don't just turn up and expect them to accommodate you.

Camping and caravanning sites have onsite electricity. Many sites have been happy to offer electric car charging for a small fee when requested.

If you want to borrow electricity in this way, take a suitable extension lead with you. Make sure that the lead also has a RCD protected plug and that the socket to plug your standard car charging cable in is protected from the elements.

Charging at work

Many G-Wiz owners make arrangements with their employers to allow them to charge up at work. Many employers are happy to provide this as it portrays the company as being environmentally friendly. In some cases, the G-Wiz owner has to pay their employer for the electricity used.

There are cases where charging your car at work causes petty jealousies with other members of staff, who see it as getting something for free that they can't have themselves.

Offering to contribute to the company for the electricity, or offering to pay a small donation to charity instead almost always resolves this problem.

Quite often, an external power socket will need to be installed in order to allow an electric car to be charged up regularly. The cost of this will vary from site to site but is rarely expensive.

Electric Car charging points

More and more towns and cities are now installing electric car charging points. Charging points are being built into retail shopping areas, car parks and at roadside parking facilities.

Local government and councils are under pressure to make electric car charging points available. The pressure is coming from politicians, environmental groups, car manufacturers and from many electricity companies. Charging point equipment is available from several manufacturers. Many new inner city developments are being designed to ensure they are electric vehicle friendly.

Businesses are also offering charging facilities for customers. Restaurants, pubs, hotels, shops and service stations are starting to offer charging facilities, often free of charge for customers and for a small fee for other owners.

In terms of electric car charging points, the United Kingdom is leading the way. There are charging facilities in dozens of towns and cities around the country, with thousands of new charging locations planned for the near future. Combined with the EV Network initiative, the United Kingdom is close to having its own nationwide charging network already in place.

Other countries are also working on their own car charging networks. Ireland, France, Spain, Portugal and Italy all have growing numbers of charging points.

Long Distance Driving

Driving for very long distances on a regular basis is not yet a practical option for any electric car, and the G-Wiz is no exception to this.

Of course, if your destination is within range of your car and you are going to be parked for several hours before returning, it may be an option to charge up your car at your destination. This effectively doubles the range of the G-Wiz.

A number of G-Wiz owners do lots of short journeys throughout a day. When added together, the overall distance travelled is significantly further than a single charge will allow them to go. In between journeys, the car is placed on charge in order to extend the range, often being charged from customer sites and creating a virtually limitless range. There are many stories of G-Wiz owners travelling over 100 miles in a single day by plugging the cars in to provide a top up charge whenever they can.

Long Term Ownership

There have been G-Wiz's on the road now for over seven years. Many G-Wiz owners have now owned electric cars for five years or more and a number are on their second or third electric car.

When asked what they like about their electric cars, many owners talk about the lack of stress when driving a G-Wiz. They are easy to drive and are much quieter than a conventional car. Several owners report that this combines to make driving a much more pleasurable and calming experience.

Many owners also talk about the cost savings of driving a G-Wiz as well as the convenience of being able to charge their cars at home and never having to go to a service station for fuel.

Range does vary with all electric cars, depending on the conditions and the age of the battery pack. In cold weather, range can decrease by as much as 20-25%. The range will also decrease as the batteries get older. The G-Wiz battery pack will require replacement every 2½-3 years (except in the L-ion model where the battery pack is expected to last 5-7 years) and as the battery pack ages you can expect the range to decrease by as much as 30%.

Yet range is hardly ever mentioned with long term owners. It simply isn't regarded as an issue. In some cases, the G-Wiz is a second car and therefore not used for long distance journeys. In other cases, the owners do not travel long distances by car at all, using the train instead. Some owners hire a car or belong to a car club for the rare times they need to travel longer distances.

The lack of a nationwide charging infrastructure is also not regarded as an issue. Only a comparatively small number of current electric car owners use the electric car charging points that already exist. Many see very little need in having a charging point network at all.

When questioned, the vast majority of people who have owned a G-Wiz for two years or longer are so pleased with them they intend to buy another electric car when they replace their existing model.

Chapter Summary

- Owning and using a G-Wiz is a new experience and it takes a little time to adapt to.
- There is a lot to like about the G-Wiz: the smooth acceleration, the lack of engine noise or vibration, the lack of noise or pollution from the car itself and the fun factor.
- New owners usually suffer from 'range fixation' which is overcome as confidence in the car increases.
- Nearly everybody forgets to plug in the car... once! Very few people forget a second time.
- There are various options available for charging up the car when travelling around. Some people use this to travel surprisingly long distances in a day.
- Most long-term G-Wiz owners are so pleased with them that they plan to buy another electric car when they replace their existing model.

Will a G-Wiz work for me?

It is important to really think about whether you are one of those people for whom an electric car is the right choice. To do this you need to consider whether it is practical in your circumstance. You will need to think about how you use a car and question what is important to you.

This process may take you some time to go through. Don't expect to have all the answers immediately. Owning an electric car is quite often a lifestyle choice and deciding whether or not it is suitable for you can take some time.

Read through this chapter, consider it, then read the rest of the book. Then come back to this chapter and re-read it when you are ready. You may come up with an entirely different set of answers the second time around.

To understand whether or not a G-Wiz will work for you, ask yourself some questions:

What benefits will I get from owning a G-Wiz?

What benefits do you expect to get from owning an electric car?

- Lower running costs?
- Better fuel economy?
- A way of allowing you to carry out your daily driving at reduced impact on the environment?
- The opportunity to drive something different, new and interesting?

You may already have some definite ideas about why you would consider owning an electric car, but you may not yet know all the benefits of owning one. Reading this book should fill in some of the gaps and prompt some fresh ideas.

Take time to consider why you want to use an electric car. There will be more than one reason. Understanding all your motivations for driving a G-Wiz will certainly help you to work out whether or not a G-Wiz is going to be suitable for you.

Where do I live and where do I drive?

If you live in a town or city and most of your driving is in built up areas, an electric car makes a lot of sense. The G-Wiz is an ideal car for the city with its compact dimensions, superb turning circle, non-polluting electric motor and nippy acceleration providing a great driving experience.

If you live in an urban or village environment, the G-Wiz may also work for you, so long as your daily range requirements are met and the lack high speed performance is not a concern.

If you do a lot of travelling on motorways or dual carriageways, a G-Wiz is not going to be the right vehicle for you. For short dual carriageway runs of a couple of miles or so, the later G-Wiz models are okay but for longer distances, the limited top speed can become an issue.

How far do I travel in an average day?

You need to ascertain whether the distance you drive in a day is comfortably within the range of a G-Wiz. If it isn't, are you able to put the car on charge between journeys?

You also need to maintain a margin for error – to ensure that you have enough range on your car to comfortably carry out all your journeys. In the United Kingdom, the average car is driven 22½ miles (36km) a day. These distances are well within the range of the G-Wiz.

However, as a general guide, you should choose an electric car with a range of around double the distance you believe you will travel between charges.

For instance, if you need to travel 20 miles (32km) a day and do not have the facility to recharge the car during the day, you need to ensure your choice of electric car has a range of at least 40 miles (64km).

Having this extra range means that you should comfortably be able to travel as far as you need each day, no matter what happens:

- In the winter, the range of the car will decrease by around 20-25% when the batteries are very cold.
- You are also more likely to require heating, lighting and windscreen wipers on at the same time, all of which have an impact on range.
- The range does decrease when the batteries are old. Ensuring you have more range than you need when you first buy your electric car ensures this does not become an issue.
- One day, you will forget to plug your car in to recharge it. If you have enough charge to use it the following day, you won't end up feeling foolish!

With its official range of 40 miles on the early cars, 48 miles on the later ones and 70 miles on the G-Wiz L-Ion, this means that if your normal daily travelling is 20-25 miles or less (or 35 miles on the G-Wiz L-Ion), you will comfortably be able to use the G-Wiz as your daily vehicle.

Can the G-Wiz work as the only car in a family?

For a few months in early 2007, I carried out an experiment to see if a G-Wiz could be the sole car for a family of four. I sold our main family car and we relied entirely on the G-Wiz. At the time, I lived in Reading and worked in Coventry – 110 miles away, and was required to travel around the country on business.

I used trains and a folding electric bike for business trips. The G-Wiz was used for ferrying children, the weekly shopping and for visiting family, all of whom lived within a 30 mile radius.

Business trips by train and bicycle worked surprisingly well. In many cases journey times were quicker than driving. I could work during my journey and the folding electric bike meant that I could continue the final part of my journey quickly and easily.

The experiment worked well for a few months and proved that it was indeed possible to just use an electric car and rely on public transport for longer journeys.

Not having a 'proper' car did mean having to plan journeys more carefully in advance and because of the amount of business travelling I did – often as much as 600 miles a week – costs were actually higher than owning a second car. For these reasons the experiment came to an end, as originally intended, after two months.

Is it big enough?

The G-Wiz is the smallest car on sale in the UK today. Whilst it is spacious enough for two adults and two young children, there is no getting away from the fact that the car is small.

Children over the age of 10 will find the rear of the car too small and whilst child booster seats will fit on the rear seats, full baby seats have to be fitted in the front of the car.

Whilst it is possible for an adult to sit in the rear seats of the car, they will certainly be cramped. It might be suitable for short occasional journeys, but you wouldn't want to be the adult sat in the back of the car for longer trips or on a more regular basis.

As well as rear luggage space, there is a small, secure storage area in the front of the car underneath the bonnet. This is large enough for a compact travel pushchair and baby bag, or for 2-3 bags of shopping.

Many owners leave the rear seats folded down and use the car as a two seater. Used this way, there is a large amount of space in the back – plenty for a week's worth of shopping for a family.

Where can I plug in?

Do you have a garage, or at least off road parking that allows you to plug your car in to charge it up at night? If you do, charging up your G-Wiz is going to be easy. You may wish to install an outside power socket but in essence, you're going to have no difficulties charging up your car.

If you park in a private communal area, then you may need to seek permission from whoever manages this parking area and will almost certainly have to pay for any work to be carried out.

Often you will get an extremely co-operative response when it comes to arranging this. If you can fit a charging point to an outside wall and there is a suitable power supply at hand, then the costs can be reasonable. If, however, you need to install a freestanding charging post and run underground cables, the cost can become very significant very quickly.

If you only have on-road parking, you may not be able to charge your electric car at your house. Some electric car owners have been known to trail cables across footpaths but this is very dangerous as it poses a significant trip hazard for young children and the elderly.

I have seen one ingenious method for roadside charging where an electric car owner ran a cable from their house into a tree at the edge of the road. They then fitted a charging socket inside the tree and cleverly disguised it all to look like a birds nest!

In some countries local council offices can arrange for an electric charging point to be installed outside your house, installing a 'power bollard' by the side of the road. Costs vary dramatically and are rarely cheap, although subsidies for installing household power bollards are being considered.

Some houses with no off-road parking do have a small yard at the front of the house. These are often not large enough for a full sized car, but the G-Wiz is often small enough to be parked on these yards. You can arrange for your local council to install a 'drop kerb' on the footpath next to your house and convert the front yard into a short driveway, thereby ensuring you always have a charging bay for your electric car.

Some G-Wiz owners have made arrangements with local businesses, allowing them to charge up at the business premises outside of business hours. There are benefits for the business in allowing this:

- There is activity at their premises outside of working hours, thereby making the property less of a target for burglary and vandalism,
- It can help local business nurture goodwill and a reputation for being environmentally friendly.

Can I charge my car elsewhere?

If your place of work has a private car park, you may wish to enquire as to whether you would be allowed to charge a G-Wiz at work, either by trailing a power lead through a window on an occasional basis, or having an external power socket installed for more regular use.

Even if you live well within range of your workplace, having the ability to charge your car at work can be useful from time to time.

Many towns and cities now have public charging points, with new ones becoming available all the time. Find out where they are and if there are any restrictions on use, even if you do not plan to use them. If nothing else, knowing where they are gives you peace of mind.

Can I remove the batteries to recharge them?

I am asked this question a lot. Many people assume the size of the batteries in the G-Wiz would be similar in size and weight to a standard starter battery used in a conventional car.

Sadly, this is not the case. It would resolve a lot of problems if they were! The batteries required to power an electric car are bulky and heavy. Imagine a battery pack at least the size and weight of a large combustion engine. You would require heavy lifting gear to remove the whole pack!

Range – the true story

The realistic range of a G-Wiz will depend on a number of factors – which model you have, the age and condition of the batteries and the ambient temperature being the three biggest factors.

G-Wiz dc-drive owners report their cars have a range of 30-35 miles with good batteries for most of the year. In the winter months, this will often drop to around 24-30 miles. With careful driving and limiting top speed to 30mph, it is possible to exceed this range and a number of owners have reported that on occasion they have been able to exceed the 40 miles official range of the car.

G-Wiz ac-drive and G-Wiz *i* owners report their cars have a range of 36-40 miles in normal day-to-day driving, reducing to around 30 miles in the winter.

To maintain the range on the G-Wiz, you should drive the car until the batteries are low on charge (so the 'fuel gauge' is close to the red zone) at least once a week. Many owners do not do this, preferring to top up the charge whenever the car is parked. This extends the lifetime of the batteries, but does mean the range that the car can travel on a single charge is reduced. Occasionally running the batteries down does ensure the G-Wiz keeps its range.

When batteries get close to the end of their useful lives – typically around the three year mark, the range of the car will deteriorate. Often, owners will continue to use their cars with a much shorter range rather than to replace their batteries when this deterioration first becomes noticeable. It is not uncommon for cars with very tired batteries to only have a range of around 10 miles.

How often do I need to be able to drive further than a G-Wiz will allow me to go?

From time to time, most people will need to travel further than their G-Wiz will allow them to go.

If you are buying a G-Wiz as a second car, this is not a major consideration. Instead you can simply use your other car for the long distance journeys.

If you are buying a G-Wiz as an only vehicle, there are options available to you:

- Join a car club that allows you access to a car as and when you need one and allows you to hire by the hour.
- Hire a car from a car hire company.
- Travel by train or other public transport.

Hiring a car for occasional use actually does make a lot of sense. It allows you to choose a suitable car for the purpose. A large car for travelling with a group of people, a small car on another occasion, even a luxury car for impressing your boss!

Sharing a car with a car club can cost as little as £3.95 per hour. This price typically includes all fuel costs and insurance. A full day's car rental can cost as little as £11.50 in the UK.

If you are regularly going to be travelling further than a G-Wiz will allow you to go and you do not have access to another car, are you going to be happy to use public transport? Is this going to be practical? Regularly commuting from one city to another by train may be a practical option for some people but may be completely impossible for others.

Road Safety and Accident Protection

In Europe, the G-Wiz is technically classed as a quadricycle, not a car. Quadricycles are lightweight four wheel vehicles which are power limited. Quadricycles are very popular in France, Belgium, Italy and The Netherlands where they have become popular for in town driving. Across Europe there are over 300,000 quadricycles registered on the road.

Quadricycles have a good safety record, mainly because they are used in large cities where most accidents are minor scrapes and driving speeds are restricted. According to the French National Interministerial Road Safety Observatory, quadricycles are three times less likely to be involved in an accident than a conventional car, whilst accidents with quadricycles are two times less likely to result in serious injuries than accidents with cars.

There are numerous anecdotes within the G-Wiz Owners' Club where owners have had minor accidents. In these cases, the G-Wiz has not sustained significant damage and the occupants have not been injured.

However, one of the main differences between a quadricycle and a car is the safety standards that the vehicles must achieve. For a car to have full European Type Approval, it has to be fitted with a driver's airbag and must undergo a series of crash tests, including a 35mph head on collision. Quadricycles do not have airbags and need not undergo crash testing.

In 2007, Top Gear magazine crash tested an early model G-Wiz dc-drive using the same tests required to pass the stringent Euro N/CAP standards. They crashed a G-Wiz at 40mph into an offset concrete block. In the tests, the safety cell collapsed. The test dummy's legs were pushed back and the steering wheel was pushed back into the dummy's abdomen.

As a consequence of these findings, REVA worked with Lotus to develop an updated version of the G-Wiz. The space frame chassis was strengthened, a redesigned collapsible steering column was fitted, crumple zones were improved and the braking system was updated, which reduced stopping distances by around one third.

The new car was crash tested at the Automotive Research Association of India (ARAI) at speeds of 25mph (40km/h) and the safety cell remained intact.

The updated car was released at the end of 2007 as the G-Wiz *i*.

What if a G-Wiz is unsuitable for me?

This is a good point to take stock. If you have already decided against buying a G-Wiz, then that is a shame but at least you have vital information. It is better than spending thousands on a car before discovering that it is not suitable for you.

There two other options that you may wish to consider:

Sharing an electric car

Do you live close to friends or family that would also be interested in owning an electric car? If so, why not pool resources and buy a G-Wiz between you? If you then share all the cars you have, you can then use an electric car for shorter journeys and another car for long distance driving.

Some communities have 'adopted' an electric car in this way. The car is available for local people to use for shorter journeys and the car is parked in a location close to a charging point.

Sharing cars seems a difficult concept to accept for many. For people in their thirties and over, the car has been seen as a symbol of freedom. Everyone from my generation can remember their first car and how they were able to use it to escape from parents and parental control.

Young people do not associate freedom with escape. To them, freedom means keeping in touch with friends via Facebook and mobile phones, or playing computer games with their friends.

To young people, freedom means sharing. Research shows that young people are far more open to the concept of sharing cars, in the same way they share video games[1].

The concept of pooling cars and sharing them between friends and family may seem alien today but is likely to become commonplace in just a few years time. Zipcar, one of the largest car clubs in the world, believe that 10% of the population in the US will adopt car sharing by 2025, whilst in the UK, over 140,000 people have now signed up with City Car Club to use their car sharing scheme.

Change your life!

This might sound drastic, but many people who buy an electric car are doing so as part of a much larger life changing transition. It is not uncommon for people to buy electric cars having recently changed jobs, retired or moved house, as part of a bigger plan to improve their lives.

Your dream life may consist of living in a cave in the side of a hillside, living in the latest ultra-high tech eco home, or just living a simpler life. Whatever it is, learn how you can reduce your dependence on your existing car first. Don't try to live your life with an electric car before you are ready. It will only lead to frustration and disappointment.

Chapter Summary

* For some people, a G-Wiz is perfect. For others, they are impractical for the time being.
* You need to consider how you use a car.
* Owning a G-Wiz is often a lifestyle choice. Don't hurry the process.
* There are a few practical points you need to consider:

    o Where you live and how far you travel.
    o Where you can plug in to recharge.
    o How often you need to travel beyond the range and how you plan to do that.

---

[1] Which World is Real? The future of virtual reality – Science Clarified.

Me and my G-Wiz

Around the world, REVA have manufactured and sold over 3,000 city cars. Approximately one third of these cars have been sold in the United Kingdom as the G-Wiz.

Owners come from all walks of life. Here are just a few of their stories.

Simon Tolbert – G-Wiz dc-drive

My G-Wiz dc-drive is a 2004 model and today has almost 30,000 miles on the clock. I am its second owner having taken delivery of the car in February, 2008. I've been using it to commute 20 miles a day to and from work, plus all the extra running around I do.

The G-Wiz is my only car. For the last fifteen years or so, almost all of my driving has been local and if I needed to travel longer distance I've always taken the train. So having a car with a range of 'only' 40 miles suited me perfectly.

Pretty much all of my driving is done in and around the Birmingham area. Occasionally I'll hit a dual carriageway, but always speed-restricted ones in built up areas. The G-Wiz does a good job of keeping up with the other traffic and is the easiest car to park. I know with absolute confidence that wherever I am going to, I can always find somewhere to park my car – even if it has to be nose-in against the pavement between two other parked cars.

Before owning a G-Wiz, I had an old Ford Fiesta. I have to say that the G-Wiz is a huge improvement in pretty much every way for the sort of driving I do.

For a start, I don't have to mess about with gears any more. Manual gears in a car that is used for commuting in city rush hour traffic is daft and I never realised quite how daft until I started driving my G-Wiz. I suffer with knee problems and have had a slight limp for a number of years. Since giving up driving a manual car and using the G-Wiz, my limp has almost completely disappeared and I am in a lot less discomfort as a result.

The fact the car is virtually silent and has no engine vibrations is a big improvement. Again, until I experienced that I never really understood how much of a difference that would make.

I use the car as a two seat car with a large luggage area and always have the rear seats folded down. On a number of occasions, we've used the car at work to visit clients. It's amazing how much computer equipment you can carry in the back of a G-Wiz! Four PCs, monitors and a couple of printers, along with the obligatory box of cables and a couple of laptops fit in without a problem and the car always provokes positive comments when you turn up at a customer site with it.

I give a lift to a neighbour every morning and evening and she also likes the G-Wiz, although it has to be said she wasn't too sure of it to start with. It was not quite her image! She is always amused by its tight turning circle and the fact it will squeeze into the tiniest of parking spaces and she likes the attention the car gets.

Because there are only a few G-Wiz's in Birmingham, the car attracts a lot of attention. Sometimes teenagers make ribald remarks about the car, but as soon as I respond with "so you haven't seen an electric car before?" they respond a lot more positively, asking questions and quite often delivering the verdict that the car is 'cool'.

When I got the car almost three years ago, it needed a brand new set of batteries. I replaced the batteries then and over 16,000 miles later the car is still going strong on the same set of batteries. I charge the batteries up both at home and at work and rarely use more than 30% of my available charge.

Friends have asked if I have seen my electricity bill jump up since switching to an electric car. I have to say I really haven't noticed any real difference. There may be a small difference here or there, but in all honesty, it isn't a big enough difference for me to have seen it. It is certainly less expensive than paying out for petrol!

Mechanically I've had the car serviced by a local garage. The total cost of maintenance, servicing and MoTs since buying the car has been £360, plus my batteries which cost me £1,400 three years ago. I estimate I've saved around £3,000 in petrol costs alone since buying my G-Wiz. I wish all my cars had been so cheap to run.

My poor car went through a rough few months in early 2009. In the heavy snows and freezing weather of January, my car got rammed by a Ford Transit van and then side-swiped by an Audi just a few days later. A few months after that, a colleague reversed into my car in the work car park.

The impact with the Ford Transit was pretty heavy, immobilising the Ford Transit and making a very large bang on impact. To my amazement, the G-Wiz stood up to the accident extremely well. My rear bumper had to be knocked back into shape (by hitting it with my fist) and I later had it remounted, and there were scuff and scratch marks, but nothing more serious than that. I was able to drive away from the accident whilst the Transit driver had to wait for a recovery truck.

In all three cases, the plastic bodywork bounced back into place and the damage was limited to scratches, scuffs and minor cracks. The car was resprayed in pearlescent white a few months later, and I treated the car to a set of new black alloy wheels and a new set of carpets at the same time. It came back looking better than new.

Would I recommend a G-Wiz? Definitely. It's a great little car to drive around in – easy to drive without having to mess about with gears. It's quiet, has no engine vibrations, can be parked anywhere and can turn on a sixpence. Would I buy another one? Absolutely.

Zarla Harriman – G-Wiz *i*

The G-Wiz seemed the obvious choice for our family. I am always very keen to embrace anything that saves our Earth's resources, and it seemed like a good bargain when it came to running costs too. My husband is a bit of a techie, and loved the idea of the G-Wiz's technology and simplicity. Our son however, thought it was OK as long as he could have a go at driving it!

Our plan was to reduce our running costs and tax/insurance bill. We looked into electric cars, and the only affordable model seemed to be the G-Wiz.

Its range and power seemed to be totally adequate for our needs, as I work in Leicester City and do about 20–25 miles per day. The range of the G-Wiz seemed to give room for those emergency detours over and above the planned ones such as the mad dash to the supermarket on the way home for cat litter!

I loved the 'cutesy' design of it, and the quirkiness of the interior. I loved being the centre of attention as I drove about, people even stopping to take photos with their phones as we went by! I LOVED driving past petrol stations, telling people how much it cost per mile, and giving people rides in it. The parking was fantastic, using the tiniest spaces. There were ALWAYS spaces free at the Leicester city centre car parks with electric car charge points.

There are not many things I disliked about this great little car. It plugged into my shed and was always full next morning, it was easy to drive and fun to own. The one worry for me was maintaining the range, so discharging properly was important. This often means that a trip out in the evening needs to be done.

The heater, however, does not do what it says on the tin…. and the draughts coming through the doors were positively freezing in winter! But in the snow, never a slip or slide – much better than any car I've ever owned!

The major problem was that it HAD to be serviced in London, and however inconvenient it was it had to go there. Borrowing a car trailer and dragging it up and down the M1 was a bind, and seemed to be a bit unnecessary. *(Note, this is no longer the case: the UK importers now offer a national servicing network)*.

I would buy another one, and what more of a recommendation could you have? However, I'd like local servicing and the lithium batteries!

Lee Ffrench – G-Wiz ac-drive

I first became interested in the G-Wiz in 2007. A lot of media coverage had been given to the car and an electric car appealed: not because I'd consider myself a 'greeny' or because I'm looking for cheap running costs – both of which you get with the car – but because it was interesting and different.

At first, I just wasn't sure I could live with the looks, size or motivation of the car. But one day in 2009 I found myself in front of the computer during the working day doing a search on an infamous auction site. I spotted my car finishing in just a few hours time. A 2007 ac-drive G-Wiz quickly became mine.

I had it brought up to Ness Point, Lowestoft to become the most easterly owned electric car in the country. At the point the car arrived I'd never seen one, never driven one and didn't even look at the car before buying it. Risky? Maybe. Fool-hardy? Probably, but like so many I'd pondered on ownership for too long and it took a bit of a gung-ho attitude to take the plunge.

So what's it like owning an electric car? Superb! You quickly grow used to the looks and the G-Wiz is very cheap to run. I've never noticed the electric on my bills. I use the *How Green is my Power* carbon calculator on www.OwningElectricCar.com to assess the output of the national grid and attempt when possible to charge it using the greenest supplies.

Servicing has amounted to a couple of tyres and a wheel cylinder. I've refuelled the car once so far with fresh batteries which should last for most of my ownership time. Driving the car is the most fun I've had behind the wheel in my motoring life. The tiny size makes squeezing through gaps you'd only consider doing on a moped and parking spaces are never a problem: there's always room for a car that's shorter than a classic Mini.

My wife has never shown that much interest in cars, but I find myself always having the 'where are you going today, what route' conversations to assess who gets the G-Wiz for the day.

So long as you have a sense of humour, a sense of fun and are OK with going over the same questions over and over again ("how far does it go mate, what speed" – and intriguingly "How much did it cost") then a G-Wiz it the perfect car for the next generation, today.

Alexander Skeaping – G-Wiz dc-drive and G-Wiz ac-drive

It was very late March 2006 when my G-Wiz dc-drive arrived. I took to it instantly. During that first 17 months of ownership I covered around 11,000 miles.

I got great service out of the car. In fact, it turned out that I had been lucky enough to take delivery of the G-Wiz with the best range performance in UK history. On one day, I actually managed over 53 miles on a single charge, which amazed everyone.

After a year of satisfied G-Wiz ownership, I'd covered around 8,000 miles with very few problems. I had the opportunity to upgrade my car to the latest model – a G-Wiz ac-drive, which promised better performance and range.

I ordered a new G-Wiz ac-drive in April 2007. I decided to treat myself to two extras: air conditioning and climate controlled seats (which can heat up on cold days, and cool down on hot ones). My new car didn't arrive until 7th August, by which time I'd already covered some 11,000 miles in my original car and had begun to notice a significant decline in overall range on the batteries, which just emphasized what a good decision I'd made to trade in.

Although I had been quite happy with my G-Wiz dc-drive, it took less than a minute after climbing into the new one before I began to appreciate what a much better car the G-Wiz ac-drive is. For a start, it is much more responsive. That was a revelation, as was the even lower noise-level. The G-Wiz dc-drive was already very quiet, but the ac-drive was virtually silent!

So, just over three years and 27,000 miles, what can I report?

If you've never driven a G-Wiz before, I think you'll be surprised at what an enjoyable motoring experience conducting a G-Wiz through urban traffic is. It's small enough to slip through any gap big enough for the average motor-bike, so you'll frequently find yourself at the front of the queue. It is amazingly manoeuvrable and squeezes into impossibly small parking bays. If all else fails there's always the option of parking sideways-on. It's so short that it's scarcely longer than the average car is wide!

The sheer simplicity of the design, with a motor capable of powering the car from standing-start to a top speed of close to 50 mph without needing a gearbox, means that this car drives like an automatic. This is particularly nice around town, as you don't have to think about anything else – except ignoring the scowls of those drivers of high-priced executive cars who started out ahead of you and end up looking at your back number-plate!

Cost-wise, there is simply no comparison. 27,000 miles of G-Wiz driving have probably cost me around £350 - £400 in green tariff electricity, compared to around £6,000 for petrol. Insuring a G-Wiz is cheap as chips. The G-Wiz is also exempt from the London Congestion Charge, which otherwise costs £8 a day. Road tax is free and in certain boroughs in London, you can park in any Pay and Display parking bay for up to four hours free of charge.

If anyone is considering buying a second hand G-Wiz – don't buy the dc-drive model because it is cheap. Instead, buy an ac-drive model because it is fundamentally a much better car. I've also driven the G-Wiz *i*, and like it even better than my ac-drive as it feels bigger inside, thanks to its curved windscreen. It also has improved brakes, is a tad faster and has a slightly longer range.

I've had few problems over those three years, mostly battery-related. In that time, I've been through two sets of batteries. The present set of batteries is the new Dynex type which come with a two-year warranty. To be honest, I've found them a big disappointment. When I first got them, we were just going into a very cold spell and, when they were new, I was getting as little as 15 miles before diving into the yellow-range on the charge-indicator. Over time, this has improved somewhat, and if I drive reasonably gently, I can get over 30 miles on a charge, sometimes even close to 40, but I've never got anywhere near the 53 miles I once got out of my original G-Wiz dc-drive. However, I'm presently sticking with the Dynex because there is a two-year warranty on them.

My only other problems have been to do with charging cables. I don't have off-street parking but I live in an end-of-terrace house, and drape a cable over the wall, encased in a black rubber cable-protector. The problems I've had have mostly been due to the repeated high current uptake weakening the plug and eventually burning through the copper wires, causing charging to cease, due to the fact that a 13-amp supply is stretched to the limit for the first hour or so of the charging cycle. Other than that, the only charging-related problems have been the three or four occasions when someone has thought to unplug the cable during the night.

I charge up mostly overnight on off-peak, using a time-switch on the mains power-point, set to come on at midnight. I also frequently plug in during the day for a boost charge whenever it looks as though I may have to sally forth later on in the day for any significant distance.

If you asked if I would go back to a petrol car, the answer is "NOT EVEN IF YOU PAID ME TO!" Granted, the G-Wiz isn't appropriate to every driver or every driving-pattern, but if your motoring predominantly involves urban or suburban driving of mostly short-ish journeys, carrying not more than two adults and two small children, then you should definitely have a serious look at this alternative, environmentally-friendly and fun way of getting around.

Purchasing and Running Costs

All electric cars are more expensive to purchase than comparable petrol or diesel powered cars. Whilst it is the cheapest electric car available to buy new in the UK today, the G-Wiz is no exception to that rule.

The offset is the running costs, which are considerably lower, especially when the cost of fuel is taken into account.

Furthermore, there is a good supply of used G-Wiz's available on the market, with prices of very early models now around the £1,000 – £1,500 mark. So if you want to own an electric car today, the G-Wiz makes the best financial sense of all the electric cars available on the market.

Road Tax

Electric cars are road tax exempt in the United Kingdom. You still need to display a road fund licence, but these are provided free of charge.

Congestion Charging

London has had congestion charging for many years. Drivers wishing to drive through the centre of London during the day have to pay a congestion charge for the privilege.

Electric cars are exempt from the London Congestion Charge. If you are travelling into the centre of London on a daily basis, this can save you over £2,000 per year.

Car Parking

In some London boroughs, along with some cities such as Milton Keynes, many car parks either offer free parking to electric vehicle users, or allow them to purchase a parking permit for a fraction of the cost of other car users.

Compared to paying for daily car parking in the centre of London, this saving alone can offset the purchase price of a new G-Wiz in less than two years.

Fuel Costs

The biggest single cost saving with an electric car is a reduction in fuel costs.

Instead of paying for fuel at a service station, an electric car owner simply plugs a cable into the car overnight and charges up using off-peak electricity.

From empty to full, the G-Wiz uses around 10 kWh of electricity (1 kWh = 1 unit of electricity on your utility bill) to provide a complete charge. In the UK, off-peak electricity typically costs around 6–8 pence per kWh.

Here's an example: If an electric car driver travels around 30 miles (48km) a day, that equates to around 900 miles (1440km) a month.

The car can be charged using off-peak electricity. 30 miles of driving with the G-Wiz would equate to around 6kWh of power.

Cost Savings

Based on 6 pence a unit off-peak cost, the cost of recharging the car would be 36 pence a night, just under £11 a month.

Compared to a conventional car that gives 35mpg (7.7km per litre, based on the imperial gallon), you're using roughly 136 litres of fuel per month to go the same distance. At a cost of £1.20 a litre, that is a total cost of £163 per month.

Based on these calculations, the cost of charging a G-Wiz is less than 7% of the cost of putting fuel into a conventional car.

The sting in the tail

These cost savings look impressive, but unfortunately they do not show the full story. It would be wrong to show the cost savings purely in terms of electricity consumption as there is an additional cost to be taken into account when charging an electric car: the cost of the batteries.

Servicing Costs

The highest costs associated with the G-Wiz are with servicing and battery replacement. The main dealers for the G-Wiz in the UK are GoinGreen who offer an excellent servicing options, including mobile 'we come to you' servicing. They offer a choice of options including annual fixed price maintenance contracts, with payments that can be spread across monthly instalments.

The biggest maintenance cost associated with a G-Wiz is the battery replacement costs. The lead acid batteries in the G-Wiz require replacement on average every 2½-3 years, although more recently, a more advanced battery has become available which is expected to last closer to four years.

Battery replacement costs vary from around £1,250 from independent specialists up to around £2,500 for the longer life Dynex batteries as fitted to the latest model G-Wiz *i*.

If you amortize the battery replacement costs over a period of several years, offsetting them against the cost of fuel in a petrol or diesel car, the overall cost is still cheaper than visiting a service station and paying for fuel every week. Unfortunately, because the battery replacement costs are so high, it can still be a painful experience when they need to be replaced.

MoT

In the UK, the G-Wiz requires an annual MoT test just like any other car. There are no special considerations for the G-Wiz and the test can be carried out by any garage with standard car testing facilities.

Resale Values

Electric car resale values have traditionally been extremely high as demand for the cars has outstripped supply. When the G-Wiz first appeared in London, demand was so high, the G-Wiz had the lowest depreciation of any new car in the UK. It easily outperformed traditionally low-depreciating brands such as Mercedes, Porsche and Audi.

Today, a 3-4 year old G-Wiz in good condition will retain around 40-50% of their original purchase price whilst a 5-6 year old model retains around 30-35% of their original purchase price.

Chapter Summary

- There is a G-Wiz available to suit every pocket – from cheap, used vehicles to brand new models with advanced lithium-ion batteries.
- Charging a G-Wiz is very cheap.
- Servicing costs are higher than other small cars, but can be made more affordable with fixed rate servicing with monthly instalment plans.
- Battery replacement costs are high, but still tend to be cheaper than putting fuel into a regular car.
- Resale values for the G-Wiz are generally very good.

Electric cars and the environment

Few subjects are more emotive than the environment and when it comes to electric cars everybody seems to have an opinion. Claims and counter claims are made and after a while it becomes difficult to identify folklore from fact.

I'm not going to use this book to jump into the climate change debate. Instead, I will do my best to explain the environmental impacts of electric cars and how they compare with conventional petrol or diesel powered cars.

This is a cut down version of the *Electric Cars and the Environment* chapter found in the book *The Electric Car Guide*. In that book, I have gone into much more detail about the environmental impacts of all forms of road transport and make reference to far more research material should you wish to investigate this subject further.

Creating a comparison between the G-Wiz and conventional cars

All electric car enthusiasts are always keen to point out the fact that their cars do not produce any pollution where they are being used. Detractors point to the coal-fired power station generating the electricity in the first place.

Both groups are making a valid point but taken in isolation both groups are not looking at the whole picture. Without looking at the bigger picture, no fair assessment of the relative merits and disadvantages of different technologies and vehicle types can be made.

The European Commission have defined a standard way for measuring the emissions from cars, based on their carbon dioxide emissions from the exhaust of the car. This is measured in grams of carbon dioxide per km travelled (CO_2 g/km).

This is known as a 'tank to wheel' measurement. A measurement of emissions from the point the fuel has been pumped into the fuel tank to the point where the energy is used.

Of course, electric cars benefit significantly from this measurement because by themselves they do not pollute at all.

However, in the same way that a 'tank to wheel' measurement does not measure the true carbon footprint of using an electric car, neither does it measure the true carbon footprint of using a conventional petrol or diesel powered car. The carbon footprint for extracting, refining and transporting the oil in the first place also needs to be taken into account.

This measurement is called a 'well to wheel' measurement and in order to be able to make a true comparison between electric cars and petrol or diesel powered cars, we need to be able to identify this well to wheel calculation for both oil and electricity.

There are several measurements that need to be considered when comparing the environmental impact of a petrol or diesel powered car with an electric car:

- Air pollution.
- How the energy (both fuel for combustion engines and electricity) is produced and transported.
- Fuel economy
- The environmental impact of batteries
- Vehicle manufacturing and recycling

I have structured this chapter to discuss these measurements separately. At the end of the chapter I then pull the separate threads together to provide a suitable comparison between petrol or diesel powered cars and electric vehicles.

Air Pollution

Whatever your opinions on climate change, there is no doubt that we suffer from over-pollution in our towns and cities.

In Germany, it is estimated that over 65,000 people die prematurely every year as a result of excessive air pollution. Across Europe air pollution reduces life expectancy by around nine months, while in some European countries the average is closer to 1-2 years[2].

Worldwide, two million people die each year as a result of excessive air pollution[3] and tens of millions of people suffer from pollution related illnesses, such as heart and lung diseases, chest pains and breathing difficulties[4] [5].

Worldwide, air pollution is now seen as a major public health issue and not just an environmental issue.

Air Pollution and Road Transport

Transportation gets a lot of criticism for creating air pollution. There is no doubt that it is one of the major contributing causes of air pollution, but it is by no means the only one. Industry and homes all create air pollution that also needs to be addressed (I discuss air pollution from power stations on page 32).

There are reasons why traffic pollution has been singled out by scientists, politicians and policy makers as the most significant pollution issue and why traffic pollution is particularly harmful:

- Worldwide, industrial and domestic pollutant sources are generally improving over time. Worldwide, traffic pollutions are becoming worse[6].
- Tiny particles within vehicle exhaust, known as particulate matters (PM), are particularly dangerous when breathed in. Particulate matters penetrate deep into the lungs and in some cases directly into the bloodstream where they have the potential to affect internal organs[7].
- Vehicle exhaust emits both volatile organic compounds (VOCs) and nitrogen oxide (NO_x). When combined with sunlight, this creates a complex chemical reaction, creating ozone. When inhaled in relatively small amounts, even by relatively healthy people, ozone can cause chest pain, coughing, shortness of breath and throat irritation.
- Carbon monoxide emissions from traffic exhaust can enter the bloodstream and reduces oxygen delivery to the body's organs and tissues.
- Toxic Organic Micro Pollutants (TOMPS), produced by the incomplete combustion of fuels comprise of a complex range of chemicals that can be highly toxic or carcinogenic. TOMPS can cause a wide range of effects, from cancer to reduced immunity to the nervous system and can interfere with child development. There is no 'threshold dose'. Even the tiniest amount can cause damage.[8]
- Traffic pollution is the number one cause of particulate matter and carbon monoxide emissions in the world and a major contributor to VOC, nitrogen oxide and TOMP emissions.[9]

---

[2]EC study – Thematic Strategy on Air Pollution, COM(2005) 446 final, 21.09.2005
[3]World Health Org: Fact Sheet No. 313 – Air Quality and Health.
[4]Air Pollution-Related Illness: Effects of Particles: Andre Nel, Department of Medicine, University of California. Published by the AAAS
[5]The Merck manual for healthcare professionals: Pulmonary Disorders.
[6]Air Pollution in the UK: 2007. AEA Energy and the Environment, commissioned by DEFRA and the Devolved Administrations, UK.
[7]EUROPA research into air quality at the European Commission MEMO/07/108, 20/03/2007
[8]Air Pollution in the UK: 2007. AEA Energy and the Environment, commissioned by DEFRA and the Devolved Administrations, UK.
[9]University of Strathclyde Energy Systems Research Unit: Environmental Pollution from Road Transport.

- In built up areas, traffic can create very high levels of pollution. This often creates visible smog that does not easily clear. This high level of pollution results in very high human exposure to these emissions.

Diesel Pollution

Emissions from diesel vehicles are particularly nasty. Carbon dioxide emissions are comparatively low, but nitrous and other particulates in the emissions have been shown to be harmful to human health and are a significant issue with local air quality in many cities.

Diesel exhaust (known as Diesel Particulate Matter – or DPM) has been shown to cause acute short term symptoms such as dizziness, headaches and nausea and breathing difficulties. Long term exposure can lead to chronic health problems such as cardiovascular disease and lung cancer.[10]

In 1998, the California Air Resources Board identified diesel particulate matter as a 'toxic air contaminate' based on its potential to cause cancer, premature death and other health problems. The American Lung Association estimates that DPM causes 4,700 premature deaths annually in nine of America's major cities.

Diesel Particulate Matter is particularly dangerous as many of the particles are very small, making them almost impossible to filter out and very easy for human lungs to absorb.

Compared to an equivalent petrol (gasoline) engine, a modern 'clean' diesel engine is likely to produce 20-30 times more nitrous emissions, as well as particulate matter which the petrol engine does not. However, both carbon dioxide and carbon monoxide levels will be lower on the diesel engine than on an equivalent petrol engine.

Average Petrol and Diesel Emissions (per litre[11]) from fuel tank to wheel[12]:

	Carbon Dioxide	Carbon Monoxide	Nitrogen Oxide	Sulphur Dioxide
Petrol	2,315g 5 pounds 1oz	140g 5oz	9.5g 3½oz	Trace[13]
Diesel	2,630g 5 pounds 13oz	237g 8½oz	37g 1$\frac{1}{3}$oz	Trace[14]

This table is based on average emissions for cars and light goods vehicles in the United States. The exact amount of carbon monoxide, sulphur dioxide and nitrogen oxide will vary from one engine design to another and will also vary depending on how efficiently the engine is running.

A word about asthma and traffic pollution

Over the past thirty years, there has been a huge rise in the number of asthma sufferers around the world. Many environmental groups claim air pollution is the cause.

However, based on a number of studies carried out in Ireland, the UK, Canada, France and the USA, the general consensus in all these countries is that there is currently little evidence that there is a definitive link between the cause of asthma and air pollution.

---

[10]University of Strathclyde Energy Systems Research Unit: Environmental Pollution from Road Transport. California Environmental Protection Agency: Air Resources Board. Diesel Health Effects.

[11]To convert these figures into emissions per gallon, multiply the figures by 3.79 for a US gallon; or by 4.55 for an Imperial gallon.

[12]Source: US National Vehicle and Fuel Emissions Laboratory / BP Fuels

[13]Sulphur-free petrol, containing less than 10 parts per million, became mandatory in 2009 across the EU. US figures may vary.

[14]Sulphur-free diesel, containing less than 10 parts per million, became mandatory in 2009 across the EU. US figures may vary.

Although air pollution may not be the cause of asthma, many people who already have asthma do report increased asthma attacks as a direct result of traffic pollution: this link between *asthma attacks* and air pollution is well known by asthma specialists.

In a recent study in Wales, 65% of asthma sufferers said they suffered coughing fits and shortness of breath as a result of traffic congestion. Research in the USA show similar figures[15].

When questioned, many asthma sufferers say they cannot walk or shop in congested areas because traffic pollution triggers breathing problems.[16]

Air Pollution and Electricity Power Stations

The amount of air pollution generated from an electricity power station varies. It is based on a number of factors, including:

- What fuel is being used to generate the electricity (coal, coke, oil, gas, biomass and so on).
- The quality of the fuel used (not all coal or oil is created alike).
- The efficiency of the power station.
- The filtration used to clean the emissions before they are released into the atmosphere.

Air pollution from electricity power stations is reducing. Even existing power stations are producing lower emissions than ever before as they switch to lower carbon fuels, install particulate and sulphur filters and become more efficient at generating electricity.

There is still a very long way to go before we have 'clean' electricity. Many countries are reliant on coal or gas for the production of most of their electricity. Moving away from these to cleaner energy sources is going to take many years.

Here are average figures for pollution for each kWh of electricity generated by different types of electricity power station:

Power Station Average Emissions (per kWh of electricity)[17]:

	Carbon Dioxide	Carbon Monoxide	Nitrogen Oxide	Sulphur Dioxide
Coal	990g 2 pounds 3oz	0.2g $< 1/100^{th}$ oz	2.8g $1/10^{th}$ oz	2.7g $1/10^{th}$ oz
Gas	400g 15oz	0.1g $< 1/200^{th}$ oz	0.4g $1/70^{th}$ oz	Trace
Oil	740g 1 pound 11oz	0.4g $1/70^{th}$ oz	3.2g $1/9^{th}$ oz	1.5g $1/20^{th}$ oz
Nuclear	16g ½ oz	Nil	Nil	Nil
Geo-thermal	122g 4½ oz	Nil	Nil	1.0g $1/28^{th}$ oz
Hydro	Nil	Nil	Nil	Nil
Wind turbine	Nil	Nil	Nil	Nil
Solar	Nil	Nil	Nil	Nil

---

[15]American Academy of Allergy, Asthma and Immunology
[16]Asthma UK opinion research, 21st September, 2007.
[17]I have various sources for this information – including discussions with representatives from EDF Energy, both in France and in the UK, E-ON and the UK National Grid, plus documented sources from the IEA Energy Technology Perspectives 2008 paper, the Laboratory for Energy and the Environment at the Massachusetts Institute of Technology and the Icelandic Energy Authority.

Average CO_2 emissions for electricity generation

Every nation has a different mix of power stations to generate electricity. The United States has a high dependency on coal, France uses nuclear power, the United Kingdom have a mix of gas, coal and nuclear and Norway and Sweden generate most of their electricity using hydro-electric power stations.

The result is there are significant differences in the average emissions for electricity production from country to country.

Emissions from National Energy Mix by Country:[18]

Country	CO_2 emissions per kWh
Canada	234g (8½ ounces)
France	87g (3 ounces)
Germany	453g (1 pound)
Iceland	138g (5 ounces)
Ireland	573g (1 pound 4½ ounces)
Norway	7g (¼ ounce)
United Kingdom	537g (1 pound 3 ounces)
United States	609g (1 pound 6 ounces)

How our energy is produced

Whether we are using gasoline, petrol, diesel or electricity to fuel our cars, we need to understand how our energy is being produced in order to understand the full 'well to wheel' energy usage of our vehicles.

How gasoline, petrol and diesel are produced

The European Union publish carbon footprint figures for burning fuels in different cars. What is not widely known is that this figure only covers the *use* of the fuel. The environmental impact of extracting the oil from the earth's crust, refining it and transporting it to the service station are not included in these figures.

Despite the huge amount of industrial processes and transportation that have to take place to clean, refine, process and move these fuels around the world, the process is actually extremely efficient. Oil refineries, for example, are around 88% efficient at converting crude oil to refined oil[19]. This is an extremely high level of efficiency for any manufactured product.

'Well to Service Station' efficiency for car fuels

According to the United States Department for Energy, the average 'oil well to service station' efficiency for car fuels is 83%[20].

This figure includes transportation of crude oil to the refinery, the refining of the oil and the transportation to the service station.

Based on this calculation, you can divide the average fuel emissions by 0.83 – or multiply by 1.205 – in order to calculate an approximate 'well to wheel' emissions of petrol, gasoline or diesel powered cars:

---

[18] Sources: International Energy Agency Data Services/Carbon Trust.
[19] Source: Estimation of Energy Efficiencies of U.S. Petroleum Refineries, Michael Wang. Center for Transportation Research. Argonne National Laboratory. March 2008.
[20] Source: DOE. Electric and Hybrid Vehicle Research, Development and Demonstration Program; Petroleum-Equivalent Fuel Economy Calculation

Average Petrol/Gasoline and Diesel CO_2 Emissions (per litre[21]) from the oil well to wheel[22]:

Fuel	CO_2 emissions per litre
Petrol/Gasoline	2,789g (6 pounds, 3½ ounces)
ULS Diesel	3,168g (7 pounds, 1 ounce)

Where our electricity comes from

As already discussed on page 32, our electricity comes from various sources and each source has a very different carbon footprint.

Managing these sources and ensuring that electricity supply always matches with demand is not a straightforward process. If supply does not match demand, the result is either a brownout or a blackout. A brownout is a voltage loss that causes lights to flicker or dim and causes electronic equipment to reset. A blackout is a complete power failure.

Electricity is delivered from the power stations to homes and businesses through power grids. These grids typically cover an entire country, often with interconnections between countries, which allows exporting and importing of power to cope with different peaks and troughs in different countries at different times.

Some power stations run constantly, while others are adjusted to cope with demand. This means that it is not always obvious where your electricity is generated from. You could live next to a coal fired power station but outside of peak times your actual electricity may actually be generated by a wind turbine half way across the country.

Your power might not even come from your own country. The United Kingdom buys electricity from France and Ireland

Canada, for example, supplies the United States with electricity and the whole of Europe now has electricity interconnections between different countries.

Demand for electricity goes up and down at different times of the day: in the United Kingdom, demand for electricity increases throughout the day, peaking mid-evening, before falling back overnight to very low levels.

Coal fired power stations

Coal fired power stations generate the highest emissions of all for generating electricity. Worldwide, approximately 41% of all electricity is generated from coal fired power stations[23], while around 49% of electricity in the United States is generated from coal.

Coal has achieved this position of dominance because of price and flexibility. Coal fired power stations can use cheap coal to generate electricity at a price that is difficult for other fuels to compete with. At the same time, coal fired power stations are reasonably flexible with their power output. When demand increases, they can boost their power generation accordingly and drop back again when demand decreases.

Coal is unlikely ever to be a truly clean fuel. It will remain a core fuel for electricity production in many countries for several years to come. Any attempt to clean the emissions from coal fired power stations as a short term measure until new power stations can be built to replace coal is to be encouraged.

---

[21]To convert these figures into emissions per gallon, multiply the figures by 3.79 for a US gallon; or by 4.55 for an Imperial gallon.
[22]Source: Discussions with representatives from BP Global and figures from the US National Vehicle and Fuel Emissions Laboratory.
[23]Source: World Coal Institute

In the United States, the CO_2 emissions from coal fired power stations averages out at 990g CO_2/kWh (2.18 pounds CO_2 per kWh). In the United Kingdom, emissions are lower because of more stringent environmental controls. Drax Power Station in North Yorkshire, for instance, has emissions of around 800g CO_2/kWh and aims to have this figure down to below 700g CO_2/kWh by the end of 2011.

Gas fired power stations

Natural gas fired power stations are the principle power generators in the UK, after significant investment by the Government in the 1980s and 1990s to move away from coal for economic reasons. In the UK, natural gas now accounts for 40% of the country's total energy needs.

Compared to coal, gas fired power stations have a significantly lower carbon footprint. In addition, the gas itself is usually piped directly to the gas station, which also reduces secondary carbon emissions as there is no transportation costs to get the fuel to the plants.

Modern gas generated electricity has a carbon footprint of around 360g CO_2/kWh (0.88 pounds of CO_2 per kWh), although some of the older gas power stations (using 'open cycle' technology) have a carbon footprint of around 479g CO_2/kWh (1 pound of CO_2 per kWh).

Oil fired power stations

Oil fired power stations are often small power stations that are used to 'top up' the power grid when there is a high demand for power. There are large oil fired power stations that are used for generating electricity from oil in the United States

They generate less pollution than coal fired power stations, but they are not as clean as gas fired power stations. They also generate more nitrogen oxide than coal fired power stations, which is regarded as a very significant greenhouse gas.

Nuclear power stations

Nuclear power stations generate very little air pollution. There is, however, radioactive waste created by nuclear power stations that has to be stored securely.

Nuclear power stations generate a constant level of energy and are often used as a 'base load' on the power grids, which are then supplemented by gas and coal fired power stations as demand rises.

France uses nuclear power stations to generate the vast majority of their electricity. Many other countries include nuclear in their general power mix.

Geo-thermal power stations

Geo-thermal energy is extracted from the heat stored deep inside the earth. Boreholes are drilled deep into the earth's crust where the temperatures reach several hundred degrees Celsius. Water is pumped in, which turns into super-heated steam. Turbines are powered by the steam which in turn generates electricity.

Geo-thermal power stations are now being built and tested in different parts of the world, including in the United States, United Kingdom and Canada.

Carbon dioxide emissions from geo-thermal power equate to around 122g CO_2/kWh (around 4½oz of CO_2 per kWh). Trace amounts of hydrogen sulphide, methane, ammonia, arsenic and mercury are also found and these are captured using emission control systems to filter this exhaust.

Hydro-electric power stations

Hydro-electricity is the most widely used form of 'renewable' energy. Once built, a hydro-electric power station produces no direct waste and has negligible emissions, although there may be environmental issues involved in the construction of hydro-electricity power stations.

20% of the world's electricity is generated by hydropower, which accounts for around 88% of all electricity from renewable sources.

Most hydro-electric power stations work by damming a river and creating a reservoir. The power comes from the potential energy of the water being driven downhill from the dam under pressure through a water turbine.

Very small scale hydro-electricity systems may also be powered by a water wheel.

Pumped Storage

The big issue with electricity production on a huge scale is that it is very difficult to store excess electricity when there is little demand and then use it when demand is at its highest. This is going to become a bigger issue in the future with the introduction of more wind turbines that generate electricity based on the strength of the wind rather than based on demand.

Pumped Storage is a way of storing this excess energy and then releasing it on demand when required. The system is based on hydro-electricity with a reservoir at the top of a hill and a second reservoir at the bottom.

During periods of excess electricity generation (typically in the middle of the night) water is pumped from the lower reservoir to the upper reservoir. During periods of high demand, water is released back into the lower reservoir through a water turbine, generating electricity.

Although Pumped Storage systems are ultimately a net user of electricity, they provide a way of storing excess energy extremely efficiently. Between 70%–85% of the electricity used to pump the water is regained when the water is released.

Like hydro-electric power stations, pumped storage has the huge benefit that they can provide electricity on demand. Quite literally, a pumped storage system can be turned on and off at a tap.

Wind Turbines

Most wind turbines rotate on a horizontal axis and consist of between three and five blades, a gearbox and an electrical generator. The turbine has to face into the wind in order to generate electricity. Small generators are pointed using a wind vane. Large generators use electric motors to move the turbine head into the wind.

The big issue with wind turbines is they only generate electricity when the wind is blowing, rather than generating electricity on demand. For this reason they are difficult to integrate into a power grid. At best they can supplement electricity from other sources rather than replace it.

Solar

With a few exceptions, solar has yet to become a serious power generator on a scale to contribute useful amounts of power to a power grid.

There are exceptions. Solar farms have been built in Germany and Spain that provide between 1–10MW of power. Other schemes are being developed in Texas, California, Mexico and Tunisia.

Solar is one of the best 'small scale' electricity generators available. If you require a relatively small amount of electricity at a location and there is no mains power available, solar is often the most cost effective and easiest energy production system to install and run.

Furthermore, quite a few electric car owners have installed their own solar power to generate electricity for running their electric car, ensuring their electric cars are truly 'green' vehicles.

Utility Grid Transmission Losses

As electricity is transmitted from the power stations to the consumers, a certain amount of energy is lost en route. These losses are reduced by transmitting electricity at very high voltages. In the United Kingdom an average of 7% of all energy generated is lost through grid transmission[24].

Electric Cars and Electricity Supply

Critics claim that the power grid cannot provide enough electricity for electric cars without significant and costly upgrading.

It is true that if every combustion engine car was taken off the road and replaced with an electric one, electricity consumption would increase. It has been estimated that across the European Union, net electricity consumption would increase by 15%[25]. Worldwide, a total transport switch to electric vehicles would increase electricity consumption to 20%[26].

However, a complete 100% shift to electric power is extremely unlikely. Neither governments nor car manufacturers are anticipating a complete switch at any time over the next 40 years. Electricity companies and governments have investigated the likely power demand based on a number of different scenarios for electric car take-up. The results make interesting reading:

- In Germany, one million electric cars travelling an average of 10,000km a year would require less than one percent of its current electricity capacity in order to provide sufficient energy[27].
- Likewise, the UK Department for Transport claims that the UK has sufficient generating capacity to cope with the uptake of electric cars, assuming a managed charging cycle targeted at off-peak periods (particularly at night) when there is surplus capacity[28].

---

[24]Investigation into Transmission Losses on UK Electricity Transmission System. National Grid Technical Report. June 2008. / Electricity Distribution Losses – a Consultancy Document. Ofgem. January 2003.
[25]The future of transport in Europe: Electricity drives cleaner! Eurelectric (2009)
[26]Energy Technology Perspectives 2008: In support of the G8 plan of action. Scenarios and Strategies to 2050. IEA (2008).
[27]The Electricification Approach to Urban Mobility and Transport. Strategy Paper. ERTRAC. 24th January 2009.

The study carried out by the Department for Transport does suggest that if significant numbers of owners started charging their cars during peak hours, significant investment may be required in the longer term.

- In the United States, the American grid could support 94 million electric vehicles (43% of all cars on the road) if they were all charged during the evening and overnight, or 158 million vehicles (73% of all cars on the road) more advanced charging techniques currently being experimented with[29].

It is widely expected that the majority of electric cars will be charged up overnight, when there is surplus capacity and this will negate the need for investment in power upgrades. Power companies are likely to promote night time charging using smart metering (see below) and discounted energy tariffs.

In the course of writing this book, I have spoken at length with power network infrastructure specialists in the United Kingdom and France about the likely impact of electric cars on the power grid. There is widespread agreement that even a significant take up of electric cars is unlikely to cause problems in the next 10-15 years. In the longer term, there is an expectation that an emergence of fast charging stations and a substantial take up of electric cars may contribute to increased peak demand for electricity by 2025-2030.

Smart Metering

Smart Meters are the next generation of electricity and gas meters, providing customers and suppliers with accurate information about the amount of energy being consumed at any time.

Smart Meters also allow energy suppliers to provide flexible tariffs to their customers so that electricity costs can be cheaper when demand is low and higher when demand is high. This offers the consumer the choice as to when they use their energy and encourages consumers to use energy as efficiently and as cost effectively as possible.

Many advanced Smart Metering systems can also be configured to switch appliances on and off depending on the cost of the energy.

In order to reduce the impact of electric cars on the power grid over the next few years, it is likely that electric car owners will be encouraged to install Smart Meters that automatically switch the cars on to charge when demand for electricity is low and therefore cheap.

Owners will be encouraged to charge their cars at night time with lower cost electricity, thereby reducing the cost to the consumer and reducing impact on the grid of a large electric car take-up.

Monitoring the emissions of your own electric car

In the meantime, it is possible for you to monitor your own emissions of your electric car by monitoring the supply and demand on the power grid yourself.

The www.OwningElectricCar.com web site includes a data feed from the UK National Grid showing the power supply mix and average carbon footprint of the electricity being generated across the UK, updated every five minutes throughout the day. The site also recommends the best times of day to plug in your electric car based on electricity demand and carbon footprint.

Data from other countries will be added to the website as the data becomes available.

---

[28]Investigation into the scope for the transport sector to switch to electric vehicles and plug-in hybrid vehicles – United Kingdom Department for Transport (2008)
[29]Electric Powertrains: Opportunities and Challenges in the US light-duty vehicle fleet. Kromer and Heywood, Laboratory for Energy and the Environment, Massachusetts Institute of Technology.

Fuel Economy

Finally, we now have a way of comparing the environmental performance for running an electric car with a petrol/gasoline or diesel powered car. We can measure the CO_2 emissions of an electric car and compare them with the CO_2 emissions of a combustion engine car.

The official fuel economy figures for the G-Wiz show that it can travel around 10km (6.2 miles) on a single kilowatt-hour of electricity. Based on the carbon footprint of the electricity used to charge up the cars, we can therefore provide a 'well to wheel' carbon footprint for running an electric car.

Across the European Union, emission figures for cars are shown as the number of grams of CO_2 generated in a single km of driving (g CO_2/km). If we use the average carbon footprint of electricity production in each country, this is what the equivalent carbon footprint would be of a G-Wiz:

G-Wiz carbon footprint

Country	g/CO₂ per km
Canada	26g
234g/kWh	1 oz
France	10g
87g/kWh	1/3 oz
Germany	50.5g
453g/kWh	2 oz

Country	g/CO₂ per km
Ireland	63.5g
573g/kWh	2¼ oz
Norway	1g
7g/kWh	1/28 oz
United Kingdom	60g
537g/kWh	2¼ oz

As you can see, there is a huge difference in CO_2 emissions depending on which country you live in. If you live in Norway where most of the electricity is generated with hydro power, your CO_2 emissions are virtually zero if you drive an electric car.

As there is wild variation between CO_2 emissions in each country, it is not possible to provide a blanket CO_2 per km measurement. It is, however, possible to create this measurement by country.

How to further improve an electric cars CO_2 footprint

In countries where there is a comprehensive mix of different power generation sources, high carbon sources of electricity production such as coal fired power stations are often put on standby when electricity demand is low. Instead, electricity is typically generated by nuclear and hydro power stations.

In the UK, where the national power grid provides detailed power information every five minutes throughout the day, it is possible to identify how much power is being produced and what the power source is, as it is being used.

Based on this information, you can quickly identify the best time to charge your electric car in order to reduce your carbon footprint. In the UK, the carbon footprint for a kilowatt-hour of electricity varies from around 290-350g/kW during the night, up to 550-600g/kW during the late afternoon.

In almost every country in the world, the carbon footprint of the electricity supply drops significantly between 11pm and 7am when demand for electricity is very low. If you can charge up your car between those times, your carbon footprint for using your electric car will be lower than if you charge up at any other time.

Carbon footprints for charging your G-Wiz at different times in the UK:

UK Average (537g/kWh)	60g/km
UK Overnight (330g/kWh)	37g/km
UK Peak early evening (600g/kWh)	67g/km

This chart shows that there is a significant difference in the carbon footprint of your car, depending on the time of day that the car is charged. If you really want to reduce the carbon impact of your electric car, the answer is to charge it up between midnight and 8 o'clock in the morning, when demand for electricity is low and the supply is produced from lower carbon sources.

You can monitor the UK power grid for yourself using the website that accompanies this book: http://www.OwningElectricCar.com

Fuel economy for combustion engine cars

By way of a comparison, here are published manufacturers figures for some of the most efficient petrol/gasoline and diesel powered cars currently on sale today, shown as 'tank to wheel' CO_2 figures.

Alongside them, I have shown an estimated 'well to wheel' figure based on the CO_2 impact of producing the fuel in the first place (based on the figures and calculations shown on page 33) by multiplying the tank to wheel figure by 1.205:

Make and Model	'Tank to Wheel' CO_2 figures	'Well to Wheel' CO_2 figures
Chevrolet Aveo	132g/km	159g/km
Fiat Panda	119g/km	143g/km
Ford Fiesta ECOtec (diesel)	98g/km	118g/km
MINI Cooper Diesel	103g/km	124g/km
Nissan Pixo 1.0	103g/km	124g/km
Smart ForTwo (diesel)	88g/km	106g/km
Smart ForTwo 52kW	103g/km	124g/km
Toyota Aygo	106g/km	127g/km
Toyota IQ 1.0 VVTi	99g/km	119g/km
Volkswagen Polo Bluemotion	99g/km	119g/km

If you compare these figures with the G-Wiz figures shown on page 39, you will see the G-Wiz provides significantly lower 'well to wheel' emissions than a conventional car.

Real world testing

As part of the research for this book, I undertook my own tests to measure real world carbon emissions of the G-Wiz and two equivalent combustion engine cars, driving on a variety of roads. You can read the results of these tests starting on page 43.

The environmental impact of batteries

It is important to measure the environmental impact of batteries (both their construction and what happens at the end of their life) when considering an electric car.

Most G-Wiz's tend to be driven much shorter distances than other cars. For the purposes of calculating the environmental impact I am going to assume that a G-Wiz has a lifespan of 12 years. I am also going to assume that during the lifetime of the vehicle, it will drive 50,000 miles (80,000km) and will require four battery packs during that time.

We can then use these figures to come up with an approximate CO_2/km figure for the environmental impact of the batteries used in the G-Wiz.

Lead acid batteries are a simple technology that is cheap and easy to manufacture. Their cases are typically made of polypropylene, the plates are made of lead and a mixture of acid and water is used as an electrolyte. Many lead acid batteries also have a very high percentage of reused materials. As a consequence, the carbon cost of manufacture is low. Around 15kg of CO_2 is generated through a build of a typical 1kWh lead acid battery.[30]

Thanks to the value of the raw materials, there is an active market in recycling lead acid batteries at the end of their lives. In Europe, virtually 100% of all lead acid batteries are recycled and in the United States, the figure is around 98%.

97% of a lead acid battery can be recycled. The casing can be melted down and reformed. Lead has a low melting point, and as a consequence it requires very little energy to be converted back into a raw material and reused.[31]

Lead acid batteries are typically recycled to produce new lead acid batteries, thereby keeping the carbon footprint down for the next generation of lead acid batteries.

Based on the G-Wiz requiring eight lead acid batteries which are then replaced every three years, the carbon footprint for all the batteries used in the G-Wiz, spread over the lifetime of the vehicle is as follows:

Estimated carbon footprint per battery:	15kg
Estimated carbon footprint per set of 8 batteries:	120kg
Estimated number of battery sets needed during the lifetime of the vehicle:	4 x 8 batteries
Carbon footprint over lifetime of vehicle:	480kg
CO_2/km based on 50,000 miles (80,000km)	6g

The environmental impact of vehicle manufacturing and distribution

In the United Kingdom, the Society for Motor Manufacturers and Traders (SMMT) estimated in 2005 the energy needed to manufacture a car in the United Kingdom translated to 600kg of CO_2, down from an estimated 1,100kg in 1999. In addition to these figures, the SMMT estimated the production of the raw materials is added a further 450kg per vehicle. It is fair to say that most industry analysts are sceptical of these figures, believing the raw material figures to be far higher.

In the United States, Honda estimates their average to be 810kg of CO_2. At time of writing, they do not have any estimates for the raw material impact.

More recently, Kia carried out a total lifecycle analysis assessment of their forthcoming Cadenza mid-range car, measuring the carbon footprint of the car from the point of manufacture to the point of destruction. The work was certified by the Korea Environmental Industry and Technology Institute (KEITI).

According to their figures, production of the raw materials for the Kia Cadenza causes the emission of 3.48 tons of carbon dioxide. A further 0.531 tons are produced in the production of the car and 0.012 tons during recycling[32]. These figures are dwarfed by the carbon dioxide emissions from actually using the car: 25.5 tons of carbon dioxide emissions based on 75,000 miles (120,000km) of driving.

G-Wiz Manufacturing

REVA, the manufacturers of the G-Wiz, is currently building a new state-of-the-environment production facility, constructed to platinum level LEED[33] Green Building certification. The site

---

[30]Source: HGI: How Green is a Lead Acid Battery? A study of the environmental impact of the lead acid battery.
[31]Source: International Lead Association: fact sheet on lead recycling.
[32]Source: Hyun-jin Cho, Sustainability Manger, Kia Motors.
[33]LEED – Leadership in Energy and Environmental Design: an internationally accepted rating system and benchmark for evaluating and certifying sustainable sites.

includes rainwater harvesting, solar energy and CO_2 monitoring facilities. Electric buses will be used to transport workers to and from work.

REVA believe their carbon footprint for an electric car will be significantly lower than the carbon footprint for other cars and not just because of their state-of-the-art factory. REVA claim electric cars use around 80% fewer parts than conventional cars, ensuring they have a very low carbon footprint.

At time of writing, they are still working to ascertain the exact carbon footprint of their vehicles and have not been able to provide me with the exact information as to their carbon footprint.

However, from mid 2011, REVA is planning to publish the carbon footprint of all their cars. Where it is not possible to reduce the carbon footprint of the car, REVA say the footprint will be offset using carbon offsetting schemes.

International Shipping

In terms of environmental impact, it makes very little difference as to whether your car was shipped around the world from the factory to the customer, or whether it was built a few miles down the road. Less than 1% of the overall carbon footprint of the average vehicle is associated with distribution[34]. International shipping makes up the smallest part of the carbon footprint associated with distribution.

A far higher part of the footprint associated with vehicle distribution is attributed to the road transportation of the car from the port to the showroom.

The environmental impact of vehicle recycling

Car manufacturers and governments alike have worked hard on ensuring that when cars come to the end of their lives, as much of it as possible can be recycled.

Across Europe, every car built since 1980 that is scrapped at the end of its life must be recycled, with a minimum of 85% of its content reused.

Car manufacturers are now legally responsible for ensuring cars they manufactured are properly recycled at the end of their life. In the vast majority of cases, well over 90% of the content of a scrapped car is now reused through vehicle recycling.

It is too early to say whether an electric car is more or less recyclable than any other type of car. It is unlikely, however, that it will be too different from existing vehicle recycling. Battery recycling is already well established and the high value of the metals used in both the batteries and the motor in an electric car will ensure it is financially profitable for vehicle recyclers to recycle electric cars properly.

Chapter Summary

Buying and running an electric car has significant environmental benefits over any other type of car:

- The carbon footprint associated with electric car production is less than other vehicles.
- You can reduce your personal carbon footprint further by charging up your car on off-peak electricity, which is typically generated from lower carbon sources.
- The carbon footprint associated with driving an electric car is less than other vehicles, even when battery replacement and recycling is taken into account.

---

[34]Ecolane Transport Consultancy: Life Cycle Assessment of Vehicle Fuels and Technologies – March 2006.

Real world economy figures

In the real world, with real world driving, it seems virtually impossible to get the fuel economy figures that the manufacturers claim for their models. Whether you're talking about combustion engine cars or electric cars, the figures appear to have little relevance in the real world other than to work as a comparison with other makes and models.

In an electric car, however, you can create your own fuel economy measurements extremely easily. Simply record your distance travelled and then measure the amount of electricity used to recharge the car using a plug in watt meter.

The test

In order to measure 'real world' economy for electric cars, I carried out a driving test along a fixed route with a G-Wiz dc-drive.

In order to provide a useful comparison, I then tested two petrol powered city cars, driving the same route in order to identify real world fuel economy and comparative carbon footprint figures for each type of vehicle.

I decided to use my own personal commute to and from work as a test route, travelling in busy traffic. The distance travelled is 7 miles (approximately 11¼km) each way. The route comprises of 2½ miles (4 km) of fast freeways and 4½ miles (7¼ km) of busy inner city roads.

The tests were carried out in and around Coventry in the United Kingdom in January 2010. Temperatures were around freezing during the whole trial. Cabin heating was used as appropriate.

Noting the temperature is important. Electric cars are less economical in cold conditions than they are in hot conditions. These tests therefore reflect a 'worse case' economy for electric cars. In warmer conditions, it would be fair to expect significantly improved figures on an electric car.

Likewise in cold conditions, combustion engine cars take longer to warm up and are also not at their most economical at the start of their journey.

Test validity

It is worth stressing that these tests have not been independently verified by any scientific establishment. Consequently, these tests can only ever be used as an indication of relative fuel economy and carbon emissions.

I have also felt that it is important that my test could be repeated by anyone else using their own cars and their own routes.

All the information and calculations I used in order to carry out my tests are included within this chapter. If a university or a scientific establishment wish to carry out similar tests in a controlled environment and would like to talk to me about my test methods, I can be contacted through the *Ask me a Question* page of www.OwningElectricCar.com.

The G-Wiz

At midnight each night, the G-Wiz dc-drive was plugged in until the battery packs were completely charged. The amount of electricity used was monitored using a watt meter and this was then multiplied by the average carbon footprint for UK electricity during the period the G-Wiz was on charge. I was charging the car overnight when the power grids are under-used and the carbon footprint averaged out at 330g/kWh.

This carbon footprint figure takes into account the carbon impact of sourcing the fuel and transporting it to the power station, the production of the power and the average transmission losses of the power as it is delivered from the power station to the car. You can view these figures yourself on the *How green is the Grid?* web page on www.OwningElectricCar.com.

I also recalculated the carbon footprint figures based on the environmental footprint of a coal-fired power station based on the Drax coal fired power station in Yorkshire – the biggest single source of carbon pollution in United Kingdom – to work out a worst case environmental footprint for the G-Wiz.

Finally, I took into account a carbon footprint for the use of the batteries, as shown on page 40.

The combustion engine cars

The petrol powered cars chosen were a brand new Toyota Aygo and a 2007 model Fiat Panda.

Both of these cars are small city cars. The manufacturers own CO_2 footprint figures show that in official tests, the Toyota Aygo produces a 'tank to wheel' footprint of 106g CO_2/km, whilst the Fiat Panda produces a 'tank to wheel' footprint of 119g CO_2/km.

These figures only reflect the 'tank to wheel' emissions, not the 'well to wheel' emissions. For our tests to be comparative to the electric car tests, well to wheel calculations have to be used.

In order to calculate the carbon footprint in real conditions, I filled the fuel tank at the start of the test. I then measured the fuel economy in litres at the end of each test by refilling the fuel tank. I calculated the CO_2 footprint based on the amount of fuel used, using the 'well to wheel' CO_2 figures shown on page 33.

Test results from the G-Wiz:

	G-Wiz
Distance Travelled	14.1 miles 22.56 km
Electricity Used	2,990 watts
Total electricity cost[35]	24p UK
Average CO_2 per kWh	330g/kWh
CO_2/km electricity usage	43.73 g/km
CO_2/km battery usage[36]	6g/km
Total CO_2/km	49.73 g/km

These figures show remarkable fuel economies and a low recharge cost for the electric cars. The carbon footprint is also low, which is helped by using off-peak electricity. Off peak electricity is much more carbon friendly than using electricity during peak times.

What about electricity from coal fired power stations?

When electric cars are powered by coal fired power stations, the carbon footprint is significantly higher than when they are powered by most other sources.

The biggest carbon polluter in the United Kingdom is Drax Power Station, emitting over 22 million tonnes of carbon dioxide per year. If a G-Wiz was charged up using electricity generated at Drax Power Station, the pollution would be equivalent to 113.11g CO_2/km. Add the 6g/km for battery usage and the total figure would be 119.11g CO_2/km.

Test results from the combustion engine cars

	Toyota Aygo	Fiat Panda
Distance Travelled	14.1 miles 22.56 km	14.1 miles 22.56 km

[35] Costs based on a night time tariff of 8p per unit in the UK.
[36] See the section on the Environmental Impact on Batteries on page 130 for a definition for this figure.

Fuel Used	1.33 litres	1.51 litres
Total Fuel Cost[37]	£1.52 UK 92¢ US	£1.73 UK $1.04 US
Official CO_2/km tank to wheel	106g/km	119g/km
Actual tank-to-wheel CO_2/km from road test tank to wheel	136.48 g/km	154.94 g/km
CO_2/km [38] well to wheel	164.46 g/km	186.70 g/km

As you can see, the carbon footprint figures that I achieved in my test are significantly higher than the official CO_2 figures. There are various reasons for this:

- The cars were driven by me and not a professional test driver. Whilst I did use eco-driving techniques, I would never claim to be the best eco-driver in the world!
- The tests started with cold engines in cold conditions.
- The cars were driven on a variety of roads, including freeways, sub-urban roads and city streets in heavy traffic conditions.
- Cabin heating was used in the cars as appropriate to keep the windscreen clear (this is also true of the G-Wiz).

Side by side analysis: well to wheel measurements

	G-Wiz	Toyota Aygo	Fiat Panda
Fuel Cost	24p UK	£1.52 UK	£1.73 UK
CO_2/km	49.73 g/km	164.46 g/km	186.70 g/km

Side by side analysis if powered by coal

If the electric cars had been powered by a coal-fired power station, the carbon footprint comparison would look like this:

	G-Wiz	Toyota Aygo	Fiat Panda
CO_2/km	119.11 g/km	164.46 g/km	186.70 g/km

Chapter Summary

- I have carried out tests on the comparative economies of the G-Wiz with two economical combustion engine cars.
- As these tests have not been independently verified, they can only ever be used as an indication of relative fuel economy and carbon emissions.
- The tests indicate that the G-Wiz is significantly better for the environment than the equivalent combustion engine car.
- Even if my electricity is generated by a coal-fired power station, my tests indicate that the G-Wiz is still better for the environment than a combustion engine car.

---

[37] Fuel price based on a UK cost of £1.14 per litre and a US cost of $2.60 per US gallon.
[38] See page 45 for information on how this has been calculated.

An electric car that powers itself?

In early 2009, REVA built a small number of cars which incorporated a solar array on the roof of the vehicle. The cars were used as part of an Indian Climate Solutions road tour when they were driven 3,500 km across India.

REVA believes that a solar powered G-Wiz could provide a range of around 3,000 km (1,800 miles) per year of sun-powered driving. The technology will see its way onto future REVA products.

Meanwhile, several G-Wiz owners have installed solar arrays on their homes in order to charge up their own cars. It still isn't cheap, but if you really want to reduce your carbon footprint running a G-Wiz purely on solar power really is a viable solution.

A final word

Electric cars are radically different and an exciting new technology that have practical uses in our daily lives and have significant environmental and economic benefits.

For many people they are the ideal vehicle, providing quiet, smooth and practical motoring for the new decade.

If, through the pages of this book, I've encouraged you to go out and try a G-Wiz, I have achieved what I set out to do. Likewise, if you have read the book and come to the conclusion that a G-Wiz is not suitable for you, this book has also served its purpose. Far better to spend a small amount of money buying a book than spend a lot of money on buying the wrong car.

Finally, if you have questions about electric cars, or suggestions on how I can improve the book, I would be delighted to hear from you. I can be contacted through the 'Ask me a Question' page on my website www.OwningElectricCar.com.

Alternatively, why not sign up to the G-Wiz Owners' Club? Membership is free and open to anyone who is interested in these cars. There is a lively online forum where you can chat with knowledgeable and enthusiastic owners. You can find us at www.G-Wiz.org.uk.

All the best,

Michael Boxwell
November 2010

CPSIA information can be obtained
at www.ICGtesting.com
Printed in the USA
379825LV00032B/1583